The *Truly* Astonishing Hypothesis

With commentary reviewing Jeff Hawkins' brilliant book *On Intelligence* and Dr. Francis Crick's *The Astonishing Hypothesis*

by John Beiswenger

Contents

Section		Page
	Forward	3
1	Dr. Crick's *Astonishing Hypothesis*	7
2	Jeff Hawkins' – *On Intelligence*	10
3	The *Truly* Astonishing Hypothesis	13
4	Three States of Existence	16
5	The Soul	19
6	Action Potentials	22
7	Memory and Consciousness	28
8	Life and Healing	35
9	Aging	44
10	Death – Transition	47
	Conclusions	53
	Bibliography	57
	Index	60
	End Notes	66
	Acknowledgements	79

Forward

I have over 40 years of experience in product research, design engineering, product development, manufacturing, product management, general management, marketing and sales in high volume consumer and commercial hard goods.[1] Half of my career was spent working for manufacturers and half as a consultant to manufacturers. I have placed over 65 products into production, monitoring pilot production in the U.S., Ireland, France, Japan, Taiwan, Hong Kong, and P.R China.

I am named on over 20 U.S. utility patents, the most recent including: color LCD touch-display technology, digital alarm clock electronics, fingerprint scanning technology, surgical instrument sterilization and bioterrorism detection technology. I am named on five International PCT patent applications.

---

[1] My resume is available at www.beiswenger.com

As a native of Milwaukee, Wisconsin, I studied at Marquette University School of Engineering and at the University of Wisconsin, Whitewater.

I started my first company to design and manufacture electronic products when I was 23, and I sold the company and the rights to my first patent a year later to an automotive accessory manufacturer where I was hired as an electronics design engineer. I joined a professional engineering firm and worked as a product research consultant to manufacturers when I was 25.

I am currently Chairman of a telehealth company which I started with three others. The company has patent-pending technology, which a registered nurse and I developed, capable of detecting respiratory infections, such as influenza, bronchitis and pneumonia, in monitored individuals before they experience significant symptoms, permitting early medical intervention. The technology can monitor individuals and entire populations, such as to control pandemics and bioterrorism attacks. The product, which is in development, is a non-invasive device used upon waking.

So, why am I qualified to critique the hypothesis and theory of Dr. Francis Crick and Jeff Hawkins? - my product research experience. Product research is the process of creating new products from scratch (starting with a blank piece of paper as I like to say). I would like to give you an example of how I used a hypothesis, during a product research project, to help solve an engineering problem.

We needed to design a touch-screen control, for a machine, using the touch-screen technology I created. The control had to display green for go, red for stop and amber for caution. It was too expensive to use a color LCD.

Using cross-aligned color polarizers to filter the light, we could get green and red. The ON state of the LCD gave us green and the OFF state gave us red. But, we asked, how can we get amber with a two-state LCD – one whose segments can just be turned on and off? "I have a hypothesis," I said.

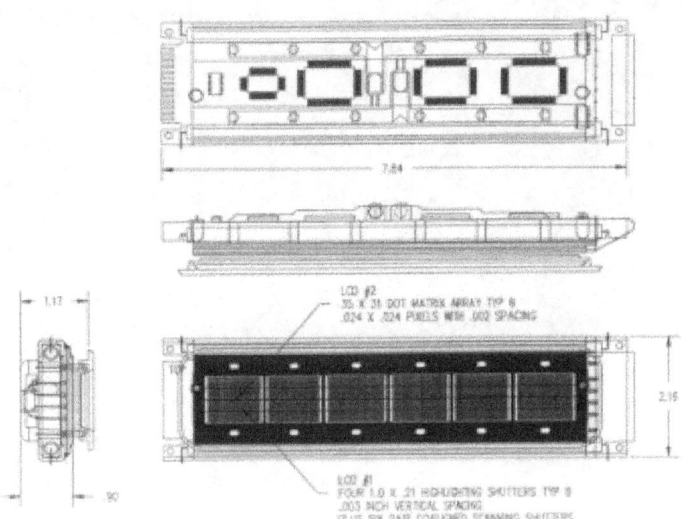

Since red and green light combined produce amber, if the LCD is dithered on and off, that is between green and red, the light coming through the LCD should appear to be amber to the human eye, assuming the LCD is dithered faster than retina neurons can react.

We designed the circuit to dither the LCD rapidly and we got the amber we needed for the control. When we

5

showed the manufacturer of the LCD a model of the control they asked, "How did you do that?"

My answer was, "by getting a very basic understanding of the related technology." That's what you have to do to create new products – get a very basic understanding of the need you are addressing and the technology you plan to employ.

That's what I did when I developed the *Truly* Astonishing Hypothesis. I first considered who we are before thinking about how our minds work. I began my study in 1983, as I will tell you later. I am also the author of three novels, *Link* (2003), *Village* (2008) and *Bridge* (2010), all based on the subject of the *Truly* Astonishing Hypothesis which I further developed during my research for the novels.

Dr. Crick's Astonishing Hypothesis

The Nobel laureate, Dr. Francis H. C. Crick, who collaborated with James D. Watson in the discovery of the molecular structure of DNA, says in his book, *The Astonishing Hypothesis - The Scientific Search for the Soul,* that, "You, your joys and sorrows, your memories and your ambitions, your sense of personal identity and free will, are in fact no more than the behavior of a vast assembly of nerve cells and their associated molecules." This is the "Astonishing Hypothesis" he offers up to the scientific community and the world to prove true with scientific facts.

In his book he goes on to say, "Most religions hold that some kind of spirit exists that persists after one's bodily death and, to some degree, embodies the essence of that human being. Without its spirit a body cannot function normally, if at all."

"As Lewis Carroll's Alice might have phrased it: 'You're nothing but a pack of neurons,'" says Dr. Crick.

Dr. Crick concludes, "If the scientific facts (gathered by the scientific community as prompted by his hypothesis) are sufficiently striking and well established, and if they support the Astonishing Hypothesis, then it will be possible to argue that the idea that man has a disembodied soul is as unnecessary as the age old idea that there was a Life Force. This," he arrogantly says, "is in head-on contradiction to the religious beliefs of billions of human beings alive today."

On the back cover a reviewer says, "In his new book, *The Astonishing Hypothesis*, Nobel laureate Francis Crick boldly straddles the line between science and spirituality by examining the soul from the standpoint of a modern scientist, basing the soul's existence and <u>function</u> on an in-depth examination of how the human brain 'sees.'" Yet the reader will find that Dr. Crick himself concludes, "By the standards of exact science, we do not yet know, even in outline, how our brains produce the vivid visual awareness that we take so much for granted." In other words, the reviewer says Crick proves his theory based on his understanding of how we are see, yet Crick himself says in his book that we have no clue as to how we see.

It would be good to note here that Jeff Hawkins agrees. He says in his book, "Our generation has access to a mountain of data about the brain, collected over hundreds of years, and the rate at which they are gathering more data is accelerating. The United States alone has thousands of neuroscientists. Yet we have no productive theories about what intelligence is or how the brain works as a whole."

Hawkins goes on to say, "Crick argued that in spite of a steady accumulation of detailed knowledge about the brain, how the brain works was still a profound mystery. Scientists," Hawkins says, "usually don't write about what they don't know, but Crick didn't care. He was the boy pointing to the emperor with no clothes. [This author thinks Crick was just being honest.] According to Crick," Hawkins continues, "neuroscience was a lot of data without a theory. His exact words were, 'what is conspicuously lacking is a broad framework of ideas.' To me," says Hawkins, "this was the British gentlemen's way of saying, 'We don't have a clue how this thing works.' It was true then, and it's still true today."

Dr. Crick only feigned his "scientific search for the soul." I propose that the scientific community search for the soul in earnest, because therein will lie the answers to visual awareness, consciousness, intelligence and man's search for meaning. Theoretical physicist Stephen William Hawking has said, "We pretty well know how the cosmos came into being, but we still don't know why." Scientists must return to considering the "why" in their search for the "how"; and the scientific discovery of the soul, its purpose and function will lead to unprecedented advances in all biological fields.

Jeff Hawkins' *On Intelligence*

Jeff Hawkins, the author of *On Intelligence*, doesn't believe in the soul either. He writes that, "Plato's solution (to understanding how we learn and apply what we've learned) was his famous Theory of Forms. He concluded that our higher minds must be tethered to some transcendent plane of super-reality, where fixed, stable ideas (Forms with a capital F) exist in timeless perfection. Our souls come from this mystical place before birth, he decided, which is where they learned about Forms in the first place. After we're born we retain latent knowledge of them. Learning and understanding happen because real-world forms remind us of the Forms to which they correspond. You are able to know about circles and dogs because they respectively trigger your soul memories of Circle and Dog."

Hawkins goes on to say, Plato's solution is "all quite loopy from a modern perspective. But if you strip away the high-flown metaphysics, you can see that he was really talking about 'invariance.'[2] His system of explanation was wildly off the mark, but his intuition that this was one of the most important questions we can ask about our own nature was a bull's-eye."

Hawkins comments about Plato's brilliant solution to how memory and learning must work, gives away Hawkins' one failure as a researcher which is his bias against the atemporal[3] part of God's creation. He has thrown the baby out with the bathwater. He missed the point. He closed his eyes to the real solution proposed by Plato.

What if I were to restate Plato's solution "from a modern perspective," which is what an open minded researcher would have done. It might read something like this:

> Plato concluded that our higher minds (the neurons of the neocortex) must be tethered to (i.e., able to communicate with) some transcendent plane of super-reality (an atemporal Particle[4] of zero mass – the fabric of the soul), where fixed, stable ideas (ancestral memories[5]) exist in timeless perfection (stored atemporally and perfectly). After we're born we retain latent knowledge of them (the ancestral memories).

"I'm an atheist," Hawkins said in an interview. "I'm not militant about it or anything like that, but I don't think you need religion to be a good guy, to be a smart guy, to do good or to be kind and caring or anything like that."

---

[2] "invariance" as used by Hawkins means an invariant representation, stored in the cortex, of an image, sound, smell or touch which is recalled and compared to the real world experience providing immediate recognition of what is being experienced.

[3] atemporal – *adj.* Timeless; outside of or apart from time. *Funk & Wagnalls New International Dictionary*

[4] "Particle" is defined later.

[5] Discussed later.

He's right. You can be a good, smart, kind and caring guy and not believe in God. My worst best friend is a designer, potter, musician and a kind, caring guy who even sings in his wife's church choir, and yet he says he doesn't believe in God. Unfortunately, because Hawkins does not believe in God, he has trouble believing in the atemporal part of reality.

In a discussion of the possibility that we live in an essentially atemporal universe, an unidentified scientist said, "There is an intuitive feeling that if we do live in a universe that lacks fundamental temporality, (this concept) might provide answers to some of the most difficult questions in quantum physics[i] and relativity." Now that's a scientist open to possibilities without bias!

Plato was on the right track and so is Hawkins. Hawkins has developed a brilliant theory of how the brain produces intelligence. I strongly recommend his book, *On Intelligence*. Plato did not have our current knowledge of science. Like Crick, Hawkins just can't get past the physical aspect of reality, because of his atheistic bias. What a shame.

The *Truly* Astonishing Hypothesis

Dr. Francis Crick is an avowed atheist. It can be assumed, therefore, that he accepts as fact that the cosmos came into existence from nothing and that man evolved from a lucky protein through natural selection over billions of years. These absurd ideas look like truth in comparison to the statement that the blueprint and biological orders for a complete, living, functioning, thinking human being are stored in the amount of DNA code contained in the zygote (a fertilized ovum). Nothing, however, is more off the mark than Dr. Crick's "Astonishing Hypothesis." It seems scientists today must mutter such absurdities while they hide in their Darwinian caves, hoping the rocks will fall on their heads rather than face God.

I too doubted the existence of God, until I came out of my intellectual hiding place to view Creation with an open mind. Once I accepted His existence, He provided a means for us to communicate. I had many questions and received many answers. The ideas that follow did not come from my mind but did pass through it. Therefore, I want to be certain that any truth found in The *Truly* Astonishing Hypothesis (hereinafter referred to as the "Hypothesis") is credited to the God and any error to me.

The Hypothesis was not originally developed as an answer to Dr. Crick's book or to answer the scientific or theological questions it appears to address. It developed from the search for logical answers to the following questions posed to me decades ago. 1. How can the single human cell, at conception (the zygote), possibly contain all of the information necessary to develop into a unique, living, thinking being? Some scientists presently reply with an absurdly incomplete answer to this question, i.e., "DNA coding." Others are not too sure.[ii] 2. How can we recall a memory of an event or place, in detail, stored in our "mind" fifty years ago or more and do it in an instant?

The *Truly* Astonishing Hypothesis was developed from the answers to two intriguing questions.

My wife, Kim, and I were on vacation at the time. I had just confirmed a rather exciting discovery of mine regarding the luteal phase of women who typically lose their pregnancies at the start of menses. I phoned in my findings to a research physician at the Medical College of Wisconsin and found that I had correctly identified their control subject as having the discovered characteristic, picking her basal temperature graph out of those of the 25 participants in the study.

When you complete a piece of work it has the effect of clearing your mind, freeing it for a new subject of investigation. Kim and I were staying in our motorhome along the Lake Michigan beach at Kohler-Andrae State Park in Wisconsin. We left the motorhome one early morning and walked along the beach. Kim said, "I wish we had brought lawn chairs so we could relax and just listen to the waves as they hit the shore." "I'll dig you a lawn chair," I said and proceeded to carve out a cool "chair" in the sand. As I labored, the two questions above circulated in my mind.

Gradually God opened my eyes to the logical answers that follow. It was only then that I began to realize how the answers applied to the Creator, creation, life, death, reproduction, healing and more that you too may uncover as you read this book.

Three States of Existence

We must start with a very basic understanding of our existence. If we don't have a grasp on who we are, then what is the purpose?

All of the hypothetical statements of the Hypothesis are in italics. That which follows some of the statements are my interpretations, explanations, references and opinions.

The first hypothetical statement may pose an insurmountable barrier to you, the reader. It is central to the Hypothesis, yet, in part, impossible to rationalize, because man is submersed in the temporal state. The figures that follow may help.

Hypothetical Statement 1: *There are three states of existence: Temporal, Sequential and Concurrent. In the Temporal State, events are spaced by time. In the Sequential State, events happen sequentially, but are not* (necessarily[6])

---

[6] I will leave my reason for adding "(necessarily)" to the discovery of

spaced by time. In the Concurrent State, all events happen concurrently.

Figure 1 is a drawing of five photographic plates that have recorded the movement of a disk from the top-left corner of the plate to the bottom-right corner. Each plate has recorded the disk at a given position as an instantaneous event. The top plate has recorded Event 1, and so forth. In between each event, represented by the space between the plates, is time (the measurement of which depends on the relative motion of the observer). This drawing represents the Temporal State.

The next drawing, Figure 2, represents the Sequential State. The same series of events is recorded, but there is no time between events.

The last drawing, Figure 3, represents the Concurrent State. The same series of events is recorded, but all have occurred concurrently. God alone exists in the Concurrent State ("I am who AM." Exodus 3:14[iii]).

It may be better to explain the three states this way. In the Temporal State, beings are limited by time[iv] and sequence. In the Sequential State, beings are limited only by sequence, and, in the Concurrent State, God is not limited by time or sequence.[v]

---

the reader.

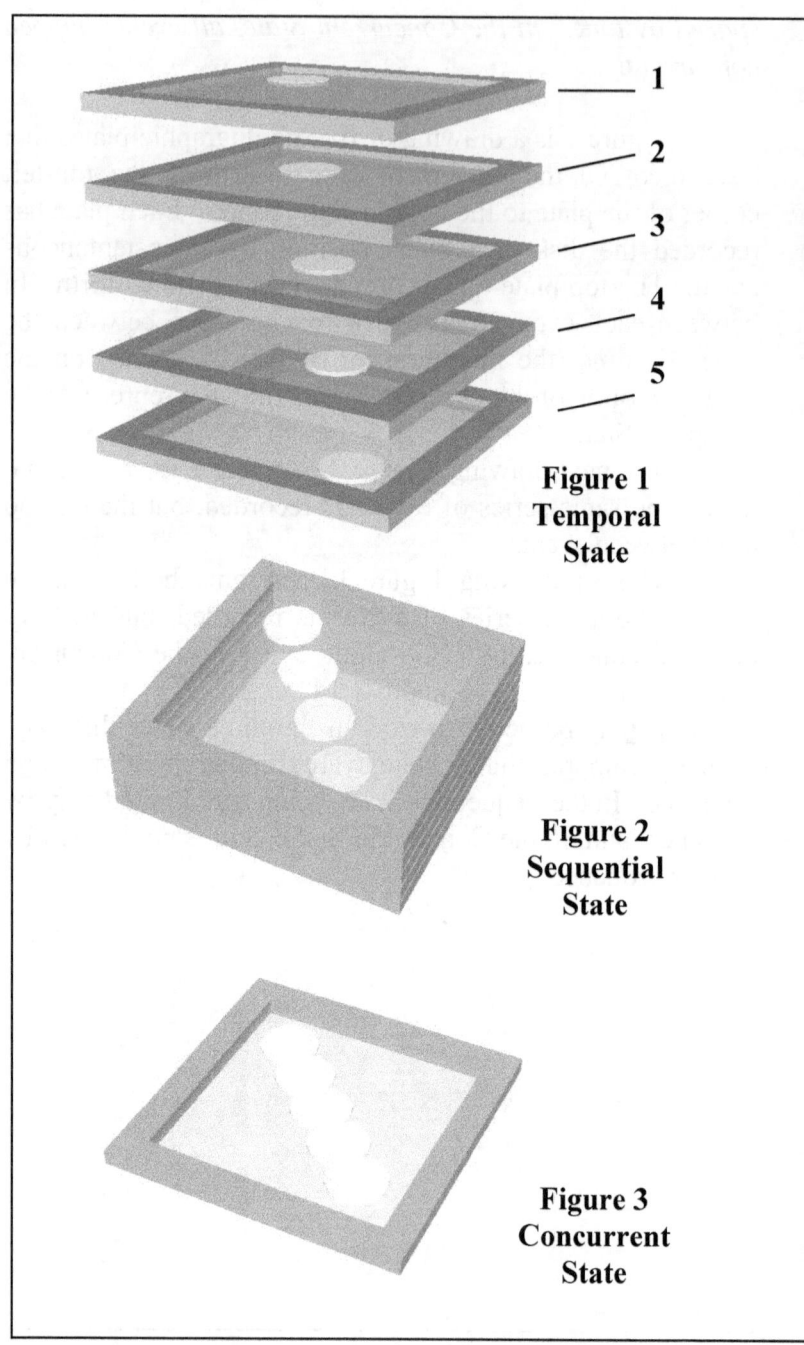

1
2
3
4
5

Figure 1
Temporal
State

Figure 2
Sequential
State

Figure 3
Concurrent
State

Section 5

The Soul

Yes, we have a soul. Crick and Hawkins don't believe that, but neither of them can provide a believable explanation of how we can store every event in our lives and recall many in an instant. Both agree, the brain is too slow, yet, Hawkins says, "You are your brain." Crick says, "You are nothing but a pack of neurons."

But, what is the soul? What is it for? What role does it play? Religious leaders have told us it is important to save our souls, but never explain just what the soul is.[vi]

Biologist Dr. David Wilcox is committed both to a strong biblical faith and to faithful, responsible science. He maintains there can be no conflict between Scripture and the natural world because God is the author of both. He is professor of biology at Eastern University in St. Davids, Pennsylvania.

Some years ago I attended a lecture by Dr. Wilcox entitled, *A Theistic Paradigm for Biology – A Biblical Perspective.* Referring to Jacob's ladder in the bible, He said, "Life is the control of the chemical (DNA), not the control by the chemical." On the way out of the classroom I walked along with Dr. Wilcox and said, "Doctor, I suggest that the soul is the source of the enormous amount of information in the cell and that DNA, figuratively, serves as a filter, adding the hereditary physical traits." He answered, "It's true the blueprint (DNA) is not the living human." I asked, "Where is the soul relative to the DNA?" He answered, "I don't know where the soul fits into all of this." He had another meeting to go to, and I headed toward the main doors of the building. I stopped when I saw him coming back toward me. He spoke as we met. "But if the soul is the source of the information, then the soul would have to be at the center of each cell." I said, "That's right." He stared at me for a moment, then turned and walked away.

Hypothetical Statement 2: A*t the functional center of every living cell is an atemporal particle of zero mass existing in the Sequential State (defining "Particle" - a biological singularity).*

Hypothetical Statement 3: *The same Particle is a component of every cell in the organism.*[7] The Particle is the "fabric of the soul."[vii]

The "Binding Problem," which obsessed Dr. Francis Crick, is defined by him as the problem of how (a set of) neurons temporarily become active as a unit. He says, "As an object seen is often also heard, smelled or felt, this binding must also occur across sensory modalities. Our experience of perceptual unity suggests that the brain in *some way* binds together, in a mutually coherent way, all those neurons actively responding to different aspects of a perceived object." Dr. Crick's brilliant mind is blinded to the

---

[7] i.e., there is only one Particle in each organism and it is at the functional center of every cell.

truth because of his choice to ignore God. All neurons (all cells) are "linked" by the Particle.

Hypothetical Statement 4:   *A state-bridging field[viii] ("Field") enables communications between the Sequential State Particle and the Temporal-State (physical) component of the cell.*

If the Field were to be withdrawn, the cell and the organism would die (defined in detail later). The Field[8] may be the "Breath of God." God sustains all life.

Hypothetical Statement 5:   *All cells are capable of receiving information from the Particle.*

Neurons (the basic structural and functional units of the nervous system) are capable of receiving a different level of information from the Sequential-State Particle and transferring it to other neurons.

Hypothetical Statement 6:   *All   neurons   are capable of transferring data to the Particle.*

Further, neurons are capable of accepting information from other neurons and transferring the information to the Particle.   The pyramidal neuron, which has a triangular shaped body (soma), are the most abundant cells of the neocortex.   In the cortex they are associated with cognitive ability.   Pyramidal neurons are the primary excitation units of the prefrontal cortex.

---

[8] It is <u>not</u> the Morphogenetic Field as defined by Rupert Sheldrake.

Action Potentials

The following reference is an excerpt from an article in the September 1992 issue of *Scientific American* entitled, "How Neurons Communicate," by Gerald D. Fischbach.

A neuron that has been excited conveys information to other neurons by generating impulses known as 'action potentials.' These signals propagate like waves down the length of the cell's single axon, and are converted to chemical signals at synapses, the contact points between neurons (i.e., contact points on dendrites leading to other neurons).

When a neuron is at rest, its external membrane maintains an electrical potential of about -70 millivolts (the inner surface is negative relative to the outer surface). At rest, the membrane is more permeable to potassium ions than to sodium ions. When the (neuron) is stimulated, the permeability to sodium increases, leading to an inrush of positive charges. This inrush triggers an impulse - a momentary reversal of the membrane potential (known as an action potential). The impulse is initiated at the junction of the cell body and axon, and is conducted away from the cell body.

When the impulse reaches the axon terminals (synapses) it induces the release of neurotransmitter molecules. Transmitters diffuse across a narrow cleft (in the synapse) and bind to receptors in the postsynaptic membrane. Such binding leads to the opening of ion channels and often, in turn, to the generation of action potentials in the postsynaptic neuron.

> **"(Action potentials) have also been traced by fine-tipped micro-electrodes positioned close enough to a (neuron) or an axon to detect the small currents generated as an action potential passes by." GDF**

Hawkins teaches, "You hear sound, see light and feel pressure, but inside your brain there isn't any fundamental difference between these types of information. An action potential is an action potential. These momentary spikes are identical regardless of what originally caused them."

The following is proposed by the *Truly* Astonishing Hypothesis: In all neurons the physical structure of DNA plays a key role in addition to the storing of coded genetic traits. In all cells[9] two turns of the double helix DNA are wrapped around bead-like nucleosomes (containing

---

[9] i.e., all cells containing DNA

histones). DNA is negatively charged, and the nucleosomes are positively charged. The assembly is twisted tightly, further coiled, folded and packed into a chromosome.

When a neuron of the brain fires, after receiving an action potential from another neuron, the pulse is coupled to its nucleus and to the DNA contained therein. The electrical characteristics of the string-like DNA wrapped around the nucleosomes produce a "disturbance" in the Field, which propagates the signal to the Particle where it is stored and instantly returned (see Hypothetical Statement 9) in the fashion of that described above.

> **The DNA/nucleosome assembly, coiled as shown in Figure 4, appears to be an extremely high frequency electronic (electromagnetic) device.**

The signal from the Particle causes a disturbance in the Field, which is propagated to the Temporal-State nucleus of the cell. There it may be translated into cellular instructions or, in the neuron, cause the neuron to fire, resulting in the transfer of the information to other neurons as described in Gerald Fischbach's article.

The Field signal, originating from the functional center of the nucleus of a cell, envelops the cell and overlaps nearby cells. In non-neural cells, it does not matter if the Field enveloping each cell overlaps adjacent cells, since the communications are generally one way. Regarding neurons, however, an overlap of sufficient signal strength could be problematic, if a disturbance in the Field enveloping one neuron is sufficient to trigger an action potential in an adjacent

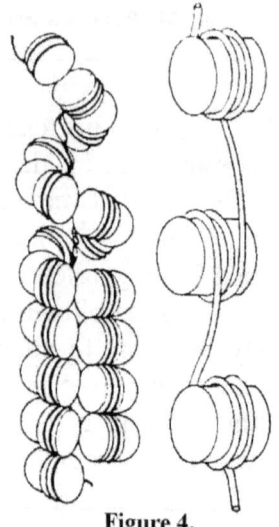

Figure 4.

neuron that may not be in the proper signal path. This is one reason why neurons are separated as they are by axons leading to synapses connected to dendrites leading to other neurons, etc.

What if, however, in the developing brain of a young child, the Field signal is strong enough or the neurons are close enough that a disturbance in the Field enveloping one neuron, under certain conditions, causes an adjacent neuron to fire, which causes another adjacent neuron to fire and so on. A signal "flashover" could occur in a part of the brain, stopped only by a break in the chain-reaction by a lesser Field signal or greater separation between neurons. This condition is called "benign childhood epilepsy." It is termed "childhood" because the child "grows out of it." The Hypothesis suggests both the cause for the condition and the reason it does not last into adulthood, both of which are presently not known.

Hypothetical Statement 7:   *Temporal-State   data communicated to the Particle is stored as Sequential-State data (defining "Parallata," pl. of Parallatum).*

Neurons of the brain receive pulsed, analog, incomprehensible information, in parallel fashion, directly or indirectly, from every nerve cell in the body and neurons in other parts of the brain.

Hawkins teaches, "The lowest of the functional regions, the primary sensory areas, are where sensory information first arrives in the cortex. These regions process the information at its rawest, most basic level. For example, visual information enters the cortex through the primary visual area, called V1 for short. Your cortex has a primary auditory area called A1 and a primary somatosensory region call S1. Eventually, sensory information passes into 'association areas,' which is the name sometimes used for the regions of the cortex that receive inputs from more than one sense. Most of these areas receive highly processed

25

input from several senses, and their functions remain unclear. The process is generally treated as though information flows in a single direction, but information in the cortex always flows in the opposite direction as well."

Hypothetical Statement 8: *The information-storage capacity of the Particle is infinite.*

Hypothetical Statement 9: *Parallata transferred to the Particle are instantly returned (reflected) to the Temporal-State nuclei of the source neurons.*

Hypothetical Statement 10: *The transfer of data between Temporal-State nuclei and the Sequential-State Particle, enabled by the Field, converts random data to comprehensible data.*

This hypothetical statement appears to address another question that puzzled Dr. Crick. All sensory analog data arriving at different parts of the brain causing neurons to fire at various rates Crick him with a few unsatisfactory answers as to how all of this activity is tied into coherent, conscious thought. On the other hand, Hawkins never actually explains how coherent, conscious thought results from the two-way flow of information he believes is always occurring.

Hypothetical Statement 11: *Sight is a joint process of the Temporal and Sequential components of visual cortex neurons.*

> **The eyes, in truth, are "the windows of the soul," for the Particle is linked directly to every one of man's senses.**

The sensor neurons of the retina receive analog information coming through the lens of the eye. Neurons of

26

the brain receive and transfer this analog information to the Particle, and it is returned to the brain in a synchronous, comprehensible manner. It is, therefore, the communication of information between the Temporal and Sequential components of the neurons of the brain, enabled by the Field, that makes vision possible.

The eyes, in truth, are the "windows of the soul." The soul (Particle), in fact, is involved in every movement man makes, every sight he sees, every smell, every sound and every touch. Man is a living being only if body and soul are communicating. Together, the source of information stored by the Particle and the neurological system of the brain form the "mind" of man, an intellectual tool controlled and focused, when man is conscious, by the soul-body unity sometimes called the "heart" of man.

Memory and Consciousness

"There is considerable evidence that our retention is much better than our normal recall would lead us to expect - indeed we may retain all of our experience." Dr. Robert C. Gilman, Ph.D., Astrophysicist.

Paul Pearsall, author of *The Heart's Code*, is a psycho-neuro-immunologist, a licensed psychologist who studies the relationship between the brain, immune system, and external factors. He has documented dozens of cases in which heart transplant recipients have received some of the memories of the donors. For example, a little girl who received the heart of a murdered child was reported to have 'recalled' the child's killer so well that she described him, and he was eventually convicted.

In his paper entitled, *Are Memories Really Stored in the Brain?* Nicholas H.E. Prince, Mathematical Physicist writes, "Essentially the thesis outlined in this paper begins at the outset by assuming that the brain itself does not store (long term) memories at all, but rather retrieves them from an external store. Indeed the implications of such a mechanism, if real, would be far reaching." He concludes, "Memories are recovered atemporally (from a timeless state)."

Ray Tillis, MD, Professor of Geriatric Medicine, UK, wrote a fascinating article in the *New Scientist* magazine about memory: "Memory is typically (viewed) as being "stored" (in the brain). But when I "remember," I explicitly reach out of the present to something that is explicitly past. (The brain is a) physical structure (knowing only the) present state. In other words, the sense of the past cannot exist in a physical (temporal) system."

Hypothetical Statement 12: *Memories are a series of Parallata stored sequentially and permanently in the Particle.*

Hypothetical Statement 13: *The Particle is able to recall memories instantly.*

Hypothetical Statement 14: *When a memory is recalled, it is returned to the same neuron-group that transferred the Parallata where it is "re-experienced."*

How often have you heard a memory recalled with, "I can almost feel the sun on my face and smell the spring air"? [ix]

Hypothetical Statement 15: *If the neuron-group that transferred the stored data to the Particle has been damaged or destroyed, and not replaced, the memory cannot, under normal conditions, be recalled* (completely, accurately or at all).

Most scientists believe that if you damage a certain area of the neocortex, memories stored in that area can be lost, which proves to their satisfaction that memories are actually stored in the brain. The above hypothetical statements suggest that memories recalled from the Particle are re-experienced through the same neuro-network that stored them. Therefore, if those networks are damaged, the memories cannot be recalled or re-experienced.

Hawkins teaches, "Even before neuroscientists were able to discern anything helpful about the circuitry of the cortex, they knew some mental functions were localized to certain regions of it. If a stroke knocks out Joe's right parietal lobe, he can lose his ability to perceive – or even conceive of – anything on the left side of his body or in the left half of space around himself. A stroke in the left frontal region known as Broca's area, by contrast, compromises his ability to use the rules of grammar although his vocabulary and his ability to understand the meaning of words are unchanged. A stroke in an area called the Fusiform gyrus can knock out the ability to recognize faces – Joe can't recognize his mother, his children or even his own face in a photograph. Deeply fascinating disorders like these gave early neuroscientists the notion that the cortex consists of many functional regions."

The Hypothesis suggests, it is the transfer of information through the Field that permits the "stacking" of Sequential data in the Particle, and it is the Field that introduces time between the synchronous Parallata returning to the neurons of the brain. If it were not for the cooperative process of the body-soul unity, consciousness would be impossible.

Just as the Hypothesis suggests, Hawkins also believes that memories are stored sequentially (in a "sequence of patterns") and can only be recalled in the same sequence; however he believes they are all stored in the cortex ("about 2 millimeters thick and, stretched flat, roughly the size of a large dinner napkin").

It is the infinite speed of the Particle that permits the amazing ability in man to recall events almost instantly from his past, in great detail (limited only by the relatively slow speed of the physical brain). It is, therefore, the Particle that makes possible the thought process and the instant comparisons (not "predictions" as Hawkins prefers) necessary for simple activities like extemporaneous speech.

Hawkins knows the brain is too slow to recall memories as fast as we can. He teaches the following: "There is a largely ignored problem with (the) brain-as-computer analogy. Neurons are quite slow compared to the transistors in a computer. A neuron 'collects' inputs from its synapses, and 'combines' these inputs together to 'decide' when to output a spike to other neurons. A typical neuron can do this and reset itself in about five milliseconds, or around two-hundred times per second. This may seem fast, but a modern silicon-based computer can do one billion operations in a second. This means a basic computer operation is five million times faster than the basic operation in your brain! That is a very, very big difference. So how is it possible that a brain could be faster and more powerful than our fastest digital computers? 'No problem,' say the brain-as-computer people. 'The brain is a parallel computer. It has billions of cells all computing at the same time. This parallelism vastly multiplies the processing power of the biological brain.'"

Hawkins continues, "I always felt this argument was a fallacy, and a simple thought experiment shows why. It is called the 'one hundred-step rule.' A human can perform multiple tasks in much less than a second. For example, I could show you a photograph and ask you to determine if there is a cat in the image. Your job would be to push a button if there is a cat, but not if you see a bear or a warthog or a turnip. This task is difficult or impossible for a computer today, yet a human can do it reliably in half a second or less. But neurons are slow, so in that half second, the information entering your brain can only traverse a chain one hundred neurons long. That is, the brain 'computes' solutions to problems like this in one hundred steps or fewer, regardless of how many total neurons might be involved.

31

From the time light enters your eye to the time you press the button, a chain no longer than one hundred neurons could be involved."

Hawkins concludes, "So how can a brain perform difficult tasks in one hundred steps that the largest parallel computer imaginable can't solve in a million or a billion steps? The answer is, the brain doesn't 'compute' the answers to problems; it retrieves the answers from memory. In essence, the answers were stored in memory a long time ago. It only takes a few steps to retrieve something from memory."

That's true, but, based on the Hypothesis, it is not the purpose of the brain itself to remember anything. The brain is the temporal component of the control and recall process. One of its highest-level roles is to facilitate "focus," allowing the being to concentrate on the most important percepts returned by the Particle. "Memory" is not stored in the brain by any neuron, neural circuit or neural network.

Hawkins, on the other hand teaches, "A typical pyramidal cell (a pyramid-shaped neuron of the neocortex) has several thousand synapses . . . then the neocortex would have roughly thirty trillion synapses altogether. That is an astronomically large number, well beyond our intuitive grasp. It is *apparently* sufficient to store all the things you can learn in a lifetime."

Really? The Hypothesis suggests that not even skills are stored in the physical brain. Nor are spinal reflexes the result of direct "wiring" between sensor and motor neurons. The Particle is involved in all memory-related activities. It takes six hours for the brain to "encode" a new skill, we are told. What is actually taking place is the construction of a neural skill circuit to by-pass the "focus function," so that the being can concentrate on other activities or on the finer points of the skilled activity in process. Spinal reflexes are produced by pre-wired focus-control bypasses, but the neurons involved in any reflex action also receive related, synchronous information from the Particle. (Note: when the

32

spinal cord above the sensor/motor neuron set is damaged, the brain cannot "modulate" the resulting action and gross reflex actions sometimes occur.)

Hawkins provides the following graphic of how information is received by the neocortex, how memories are retrieved and how information is <u>compared</u> to current experiences. I have added the true repository of all memory, the Particle., as suggested by the Hypothesis.

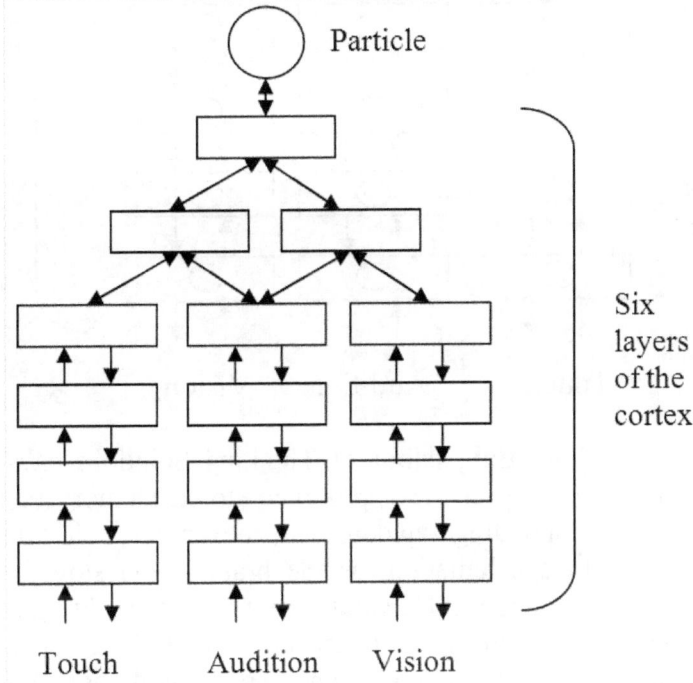

The upward arrows represent data arriving at random from all sensor neurons, incomprehensible to the brain. The data is stored in the Particle, sequentially, and returned as pulsed, comprehensible percepts[10].

---

[10] Percept – In clinical psychology, a single unit of perceptual report. Author definition: a singular, instantaneous, comprehensible perception of an experience including reports from all senses.

It is more likely that the Particle is involved in the entire process at every layer of the cortex.

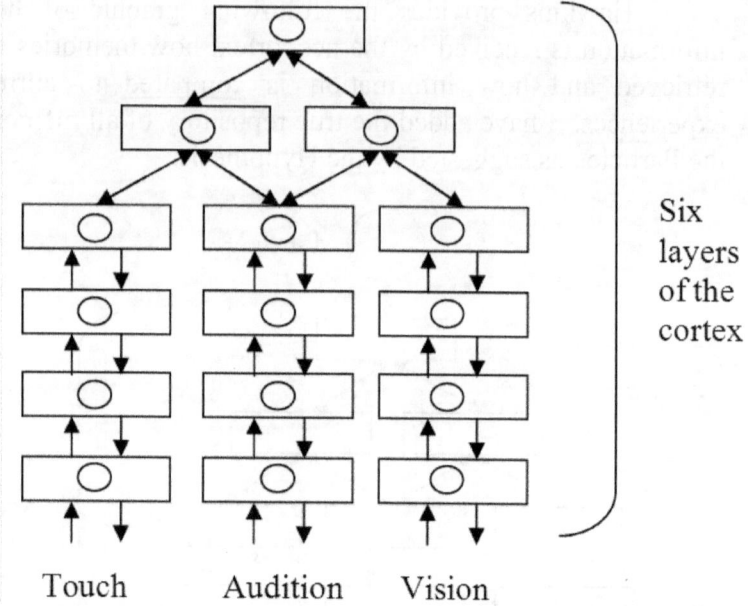

Six layers of the cortex

Touch   Audition   Vision

John Ball, editor of Thinking Solutions says, "The brain must select the appropriate stored memory to trigger the corresponding rapid muscle movements. It cannot do this with a calculation, as the brain is too slow." That statement supports Hawkins' theory. I will add that, if left up to the brain alone, a dancer could not dance, a tennis player could not play, and you could not carry on extemporaneous discussions. The brain is just too slow and scientists know it.

Life and Healing

"I am unable to believe that any machine (referencing the cell) can be designed that contains an instruction library (DNA) which anticipates all the mishaps and glitches of a billion years of evolution without crashing over and over again." Guenter Albrecht-Buehler, Professor of Cell and Molecular Biology at Northwestern University.

"A set of blueprints is not a house; the DNA of a zygote is not a human being." Garrett Hardin, professor of biology at the University of California at Santa Barbara.

Hypothetical Statement 16: *The Particle contains the source of all necessary information relating to the cell including an encoded, perfect likeness ("Likeness") of the living organism to which the cell belongs.*

Do not confuse the encoded Likeness with the imaginary "homunculus" Dr. Crick refers to.[x] Many biochemists today suggest that the genome is the encoded

35

"blueprint" of a human being. The Hypothesis is just suggesting it takes much, much more than DNA to "encode" a human being.

Hypothetical Statement 17: *Information from the Particle regarding the Likeness is filtered, figuratively, through the hereditary traits encoded by the structure of DNA in the nucleus of the cell.*

Temporal factors, such as the physical health of the cell and other environmental factors, can affect the communications between the Particle and Temporal components of the cell.

Although the Particle is indivisible, it can be united with another Particle to form a new and unique Particle.

Hypothetical Statement 18: *At the true moment of conception, the Particles of both male and female unite, creating a new, unique Particle incorporating a perfect Likeness of a new and unique body.*

Hypothetical Statement 19: *Concurrently at conception, the building blocks of DNA from both male and female fuse into the chromosomes of a new and unique cell (zygote) as controlled by the new Particle.*

Hypothetical Statement 20: *The shared Field gives life to the new cell and cell division begins, as guided by the organism's Particle* (Items 18, 19 and 20 define procreation - "with the help of God." Genesis 4).

The Hypothesis leaves little doubt that from the true moment of conception a new being is alive and growing.

Excerpts from:

"Stem Cells: A Primer"

Human development begins when a sperm fertilizes an egg and creates a single cell that has the potential to form an entire organism. This fertilized egg is **totipotent**, meaning that its potential is total. In the first hours after fertilization, this cell divides into identical totipotent cells. This means that either one of these cells, if placed into a woman's uterus, has the potential to develop into a fetus. In fact, identical twins develop when two totipotent cells separate and develop into two individual, genetically identical human beings. Approximately four days after fertilization and after several cycles of cell division, these totipotent cells begin to specialize, forming a hollow sphere of cells, called a blastocyst. The blastocyst has an outer layer of cells and inside the hollow sphere, there is a cluster of cells called the inner cell mass.

The outer layer of cells will go on to form the placenta and other supporting tissues needed for fetal development in the uterus. The inner cell mass cells will go on to form virtually all of the tissues of the human body. Although the inner cell mass cells can form virtually every type of cell found in the human body, they cannot form an organism because they are unable to give rise to the placenta and supporting tissues necessary for development in the human uterus. The inner cell mass cells are **pluripotent** - they can give rise to many types of cells but not all types of cells necessary for fetal development. Because their potential is not total, they are not totipotent and they are not embryos. In fact, if an inner cell mass cell were placed into a woman's uterus, it would not develop into a fetus.

The pluripotent stem cells undergo further specialization into stem cells that are committed to give rise to cells that have a particular function. Examples of this include blood stem cells which give rise to red blood cells, white blood cells and platelets; and skin stem cells that give rise to the various types of skin cells. These more specialized stem cells are called **multipotent**.

A primary goal of (stem cell research) would be the identification of the factors involved in the cellular decision-making process that results in cell specialization. We know that turning genes on and off is central to this process, but we do not know much about these "decision-making" genes or what turns them on or off. Some of our most serious medical conditions, such as cancer and birth defects, are due to abnormal cell specialization and cell division. A better understanding of normal cell processes will allow us to further delineate the fundamental errors that cause these often deadly illnesses.

Source: NATIONAL INSTITUTES OF HEALTH - May 2000

Hypothetical Statement 21:   *Growth and healing*[xi] *are the same process and are similarly directed by the Particle.*

Hypothetical Statement 22:   *When an organism is injured or diseased, and cells are damaged or destroyed, it is the Particle that provides the correct code (Likeness) with which to direct the physical regeneration process when possible.*

Each living cell receives specific instructions from the Particle, including instructions to divide, when necessary, for the growth or healing of the organism.   If a cell is missing, diseased or sufficiently damaged, it cannot receive these instructions, nor can it in any way report its condition. Instructions to divide for the growth or healing of the organism must be given to a healthy cell adjacent to or in line with (heretofore defines "Corresponding") the position of a missing, diseased or damaged cell.

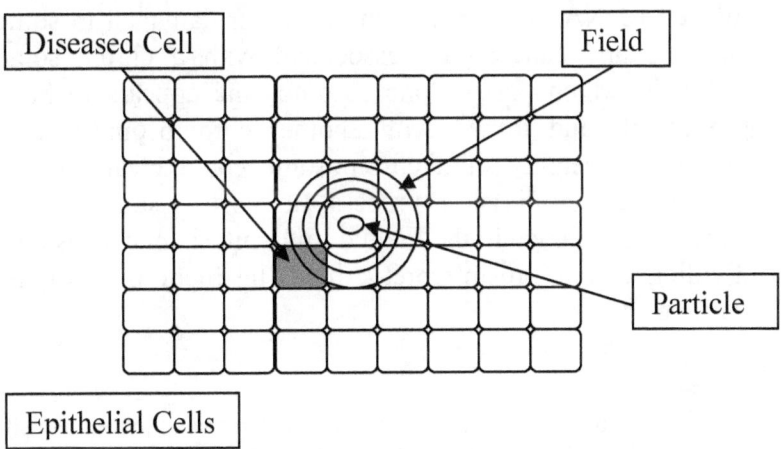

Diseased Cell   Field

Particle

Epithelial Cells

The Field, carrying the instructions from the Particle to the cell, radiates from the cell, reaching the nuclei of Corresponding cells.   There the cell-specific information is forwarded (reflected) by the Corresponding-cell nuclei (i.e., the structure of the chromosomes and DNA) to the Particle. Non-reporting cell positions are identified in this manner. When the need for cell division is recognized, and a divide instruction is sent, other Corresponding cells are inhibited from dividing by the cell-specific instructions radiating from the dividing cell. In this way the growth or healing process is controlled.[xii]

When a cell receives instructions from the Particle, and existing, even healthy Corresponding cells do not, for any reason, forward the cell-specific instructions carried by the Field; the originating cell may be instructed to divide even though there is no need for division. Since a cancer cell does not recognize proper instructions regarding shape, purpose, color, structure, location or division limits, it can be assumed it may also be incapable of being a source for correct cell-specific instructions to Corresponding cells. Thus no Corresponding cells are recognized by the Particle, and the cancer cell continues to receive instructions to divide.[11]

Although the DNA code of a cancer cell remains identical to a healthy cell, the physical structure of some of the cell's DNA may have been altered or damaged in some way. Many cancers are associated with a chromosome defect in which part of one chromosome appears to have broken off and joined with another chromosome. This improper assembly of a chromosome could clearly occur during the cell-division process (mitosis – see previous page). Interference in the Field during mitosis may cause the dividing cell to misinterpret critical instructions from the Particle.[xiii]

---

[11] It can also be assumed that cancer cells Corresponding to healthy cells are incapable of receiving and reflecting the Particle cell-specific information from the healthy cells, and are thereby identified by the Particle as non-reporting cell locations, causing even the healthy cells to divide. This could multiply the rate of unwarranted growth in the area of the cancer cell.

During mitosis it appears to be the centrosome that is receiving cell division instructions from the Particle, since the structure of DNA is disassembled during the process.

"The centrosome consists of two structures called

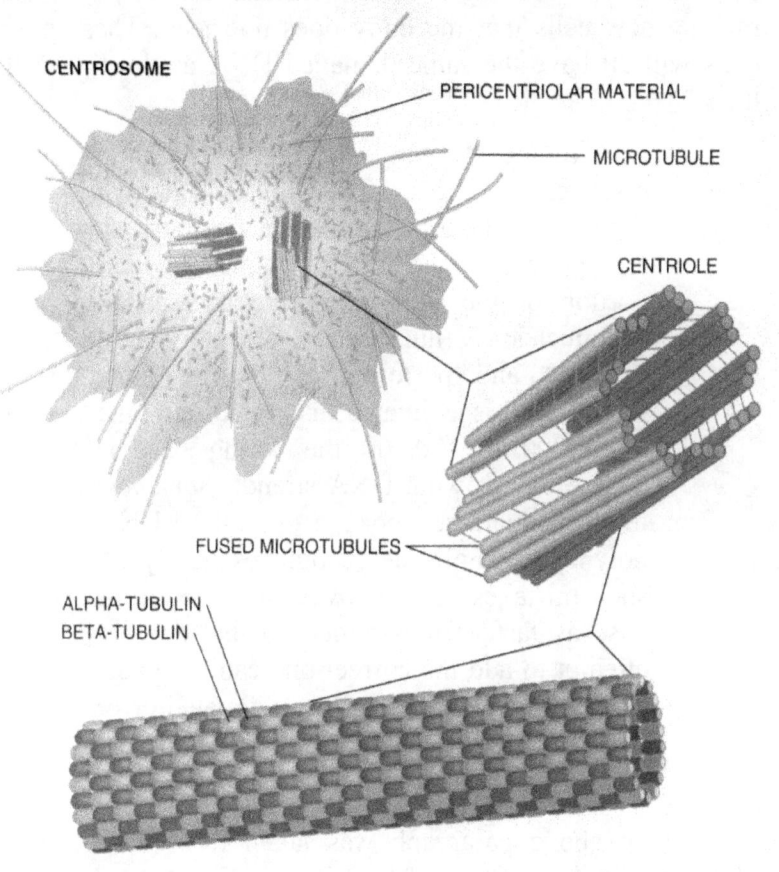

CENTROSOME

PERICENTRIOLAR MATERIAL

MICROTUBULE

CENTRIOLE

FUSED MICROTUBULES

ALPHA-TUBULIN
BETA-TUBULIN

centrioles set at right angles to each other and are surrounded by a cloud of pericentriolar material. Seen in cross section, a centriole reveals a pinwheel structure made of structural elements called microtubules (Scientific American, June 1993 – David M. Glover)."

The American Cancer Society says, "Cells become cancer cells because of damage to DNA. DNA is in every cell and directs[12] all its actions. In a normal cell, when DNA gets damaged the cell either repairs the damage or the cell dies. In cancer cells, the damaged DNA is not repaired, but the cell doesn't die like it should. Instead, this cell goes on making new cells that the body does not need. These new cells will all have the same damaged DNA as the first cell does."

> "In recent years one of the intriguing discoveries has been that while one part of the DNA polymerase molecule functions as a polymerase, attaching nucleotides, another portion of the same molecule acts as an "exonuclease" (nucleotide - cutting - out enzyme) and performs a "proofreading" function. It is estimated that about one time in 10,000 to 100,000 the wrong base is added to a growing DNA strand. *Somehow* the exonuclease portion of the DNA polymerase molecule recognizes nearly all such mistakes and removes each erroneous base as fast as it is added so that another attempt to add the correct one can be made. The result is that there is an estimated error rate of only one in one billion base pairs copied (during mitosis)."

The above paragraph was taken from the *Biology Coloring Book* by Robert D. Griffin. The "*somehow*" is explained by the Hypothesis. Based on the Hypothesis, if the communications between the Particle and the cell were perfect, no errors would result (or remain). Gross mistakes causing the displacement of part of a chromosome could also

---

[12] DNA is the chemical formula, but cannot by itself "direct" cell actions.

result in repeating errors in reception of information from the Particle through the Field. Such flaws would also affect the ability of cells to "reflect" cell-specific information back to the Particle as described in later paragraphs.[13]

Cell division is normally inhibited if nutrient levels are below certain limits. Cancer cells (and possibly healthy adjacent cells), on the other hand, will divide even when nutrient levels are one tenth of this limit. The reason may be that the Particle recognizes the non-reporting cell positions as an injury to the organism, not just normal cell replacement as in the case of surface skin cells.

Though the strength of the Field signal may drop in direct proportion to the square of the distance from the functional center of the nucleus from which it is originating, the Field reaches more than just the Corresponding cells or even cells of the same type. In this manner the Particle receives an encoded "picture" of the entire organism. So it is with the epithelial cells of the skin. The Particle receives a precise thickness measurement when it sends cell-specific information to a germinating layer cell through the Field, which radiates to (and probably beyond) the unattached surface layer cells.

If a missing, diseased or damaged cell is beyond the physical range of the Field radiating from a healthy cell, it is not recognized as a non-reporting cell location because no cell is expected within the range covered. Through action potentials, neurons are capable of transferring information from other neurons to the Particle by means of the Field and from the Particle to other neurons. In part, neurons are spaced to avoid causing an action potential in an adjacent neuron when its neighboring neuron generates a data-carrying disturbance in the Field, which also overlaps the adjacent neuron. This requirement makes the strength of the radiating Field more critical with regard to neurons,

---

[13] Do some wavelengths of Electromagnetic Radiation interfere with the Field? Probably.

especially in adults, when it comes to the process of recognizing adjacent, damaged, diseased or missing neurons.

The tree-like structure of cells forming the central nervous system, versus the layered structure of epithelial cells, further complicates cell replacement, and the healing of large gaps in the structure from injury, disease or other causes of cell death would seem to require a different process, guided by the Particle. However, when new neurons are surgically deposited in these gaps the cell replacement process described above would be facilitated.

[Note: Perhaps the body needs sleep because the thought process (consciousness) occupies the temporal/ atemporal "link" during waking hours to the extent that non-REM sleep periods are required for communications involving cell functions and healing. Perhaps, also, in between these regeneration communications the "system" releases the mind temporarily, which begins to dream. [xiv]]

Aging

We were designed to live forever in health and happiness. How could it be any other way?

Leonard Hayflick, PhD, a professor of anatomy at the University of California, San Francisco, says, "The accumulation of new insights has made it possible, for the first time, to understand the biological reasons for the aging of animals and humans. Aging occurs because the complex biological molecules of which we are all composed become dysfunctional over time as the energy necessary to keep them structurally sound diminishes. Thus, our molecules must be repaired or replaced frequently by our own extensive repair systems," Hayflick said. "These repair systems, which are also composed of complex molecules," he explained, "eventually suffer the same molecular dysfunction. The time when the balance shifts in favor of the accumulation of dysfunctional molecules is determined by natural selection and leads to the manifestation of age changes that we recognize are characteristic of an old person or animal. It must occur after both reach reproductive maturity, otherwise the species would vanish." Hayflick also noted that these repair and maintenance systems are called "determinants of longevity," which is a phenomenon different from the aging process itself. "These fundamental molecular dysfunctional events lead to an increase in vulnerability to age-associated disease," he said. "Therefore, the study, and even the resolution of age-associated diseases, will tell us little about the fundamental processes of aging."

Hypothetical Statement 23: *Man's Temporal body was designed to live forever. It is the imperfect communications between the Particle and cells that gradually bring aging to the organism.*

"Aging occurs on the cellular level in a number of ways. The strands of DNA that guide a cell's physiological process [through instructions from the Particle] can be damaged during the normal function of generating RNA, the cell's messengers. The presence of repair mechanisms [the polymerase molecule directed by the Particle] substantiates this notion, because if the DNA was not properly repaired, it would impair cellular functions. Thus, it is thought that by either wear and tear or through improper repair over time the DNA could be destroyed and the cellular reproduction function impaired." (*Grolier Multimedia Encyclopedia*)

In an article by astronomer Dr. Hugh Ross, he writes, "Medical experts agree that cosmic radiation plays a significant role in limiting human life spans. Astronomers agree that the vast majority of this life-limiting radiation comes from supernovae, cataclysmic explosions of super giant stars." The article goes on to say that astronomers Erlykin and Wolfendale confirmed that the Vela supernova is indeed the prime contributor.

The Hypothesis confirms much of what is known about wellness and long life. Adequate sleep is necessary. During non-REM sleep communications between the

Particle and cells are uninterrupted by conscious thought which may fully occupy the "system" during waking hours. Eating healthy foods maintains healthy cells and provides the chemical components for cell reproduction. Undernourished cells, including neurons, may not be as able to receive, decipher and carry out instructions from the Particle relative to the Likeness. Neurons may not be able to produce disturbances or respond as quickly (or fully) to disturbances in the Field produced by the Particle, therefore cognitive powers and memory-recall will be affected.

Avoiding radiation takes on a whole new meaning when the Field is taken into consideration. We know that ultraviolet radiation damages the DNA structure of cells. Instructions from the Particle will be either not received, will be misinterpreted or not acted upon. Dr. Michael Reacholi, manager of the WHO's Electromagnetic Fields Project, told a news conference that, "There are key issues that still need to be resolved because there have been suggestions that electromagnetic fields may produce cancers or memory loss or other neuro-degenerative diseases." The Hypothesis suggests, using a cellular phone with its antenna close to the brain is probably not a good thing to do. Living near low-level electromagnetic radiation (EMR) from power lines over a long period of time may not be safe. The Hypothesis also suggests that EMR could disrupt communications through the Field. However, there will be no direct cause/effect found until the existence of the Field is recognized.

Death - Transition

Neither Crick nor Hawkins say anything about death. In the index of their books, death is not listed. Why not? Death, or "transition" as I prefer to call the process, is a natural part of our existence. There's much interest. A search on Altavista.com for articles on death returned 1.98 billion results.

We can't talk about transition (physical death) without discussing near-death experiences. Dr. Michael Sabom is a cardiologist in private practice who has studied near-death-experiences for over 20 years. He reports as a scientist about what he has learned by interviewing nearly 50 individuals who "returned from death's door."

Dr. Sabom's latest book, *Light and Death*, shares with the world his findings. Sabom, also now a born-again Christian, scrutinizes near-death-experiences in light of what the Bible has to say about death and dying, the realities of light and darkness, and the gospel of Jesus Christ.

In his book, Dr. Sabom relates an interview with a

surgery patient he named "Pam Reynolds," a 35-year old woman, who said she was operated on for a giant artery aneurysm in her brain. She told him she underwent a dangerous surgical procedure nicknamed, "Standstill," because they cooled her body to 60 degrees, stopped her heart and drained all of the blood from her brain. She was by all clinical standards, quite dead. During that time, Pam had an Out-of-Body experience which is available in the End Notes section.[xv]

When Dr. Michael Sabom heard Pam describe the bone saw that the surgeon used to open her skull, that is "The 'saw' thing . . . looked like an electric toothbrush," he said "No way," and filed away the interview tape. A year later he decided to research her story. He called the company that made such saws and they sent him a student's user manual with pictures. He wrote, "I was shocked by the accuracy of Pam's description of the bone saw as an electric toothbrush and with the 'socket wrench case,' in which the equipment is kept." However, she said there was a groove at the top of the saw and the picture I had showed no such groove. Another detail of the saw blade matched her story.

Pam also said she heard the cardiologist say that her veins were small. Sabom contacted Pam's surgeon who told him it was the cardiovascular surgeon who commented about the small veins in her report, and that Pam could not have heard her speak because Pam's ears were covered with tightly fitting earphones that produced a loud clicking noise and blocked all outside sound. None-the-less, Pam's description of what was said was accurate.

P.H.M. Atwater carefully reviewed many near death experiences and found a similarity that could not be ignored. Here is what she found:

- Most felt a sensation of floating out of one's body . . . where all that goes on around the vacated body is both seen and heard accurately (like Pam Reynolds reported).

- Passing through a dark tunnel or dark hole or encountering some kind of darkness. This is often accompanied by a sensation of moving or acceleration.
- Ascending toward a light at the end of the darkness; a light of incredible brilliance, with the possibility of seeing people, animals, plants, lush outdoors and even cities within the light.
- Greeted by friendly voices, people or beings who may be strangers, loved ones or religious figures. Conversation can ensue; information or messages may be given.
- Seeing a panoramic view of the life just lived, from birth to death or in reverse order.
- A reluctance to return . . . but invariably realizing their job on earth was not finished or a mission must yet be accomplished before they can return to stay.
- A warped sense of time and space; discovering time does not exist.
- And some experience a disappointment at being returned.

Now let's take our study to the next step. What happens when we die. Do we die? What does the Christian Bible have to say about death. Keep the words and phrases I have underlined in the following verses in mind as you read the hypothetical statements.

Jesus said, "I am the resurrection and the life. Anyone who believes in me will live, even though they die; and whoever lives by believing in me will never die." John 11:25 (TNIV)

1 Thessalonians 5:23 (TNIV) "May God himself, the God of peace, sanctify you through and through. May your whole spirit, soul and body be kept blameless at the coming of our Lord Jesus Christ."

John 11:11-14 (TNIV). "After he had said this, he went on to tell them, 'Our friend Lazarus has fallen asleep, but I am going there to wake him up.' His disciples replied, 'Lord, if he sleeps, he will get better.' Jesus was speaking of his <u>death</u>, but his disciples thought he meant <u>natural</u> sleep. So then he told them plainly, 'Lazarus is dead.'

When they arrived at the tomb, Jesus looked up and said, 'Father, I thank you that you have heard me. I know that you always hear me, but I said this for the benefit of the people standing here, that they may believe that you sent me.' When he had said this, Jesus called out in a loud voice, 'Lazarus, come out.' The <u>dead</u> man came out, his hands and feet wrapped with strips of linen, and a cloth around his face." John 11:41-43 (TNIV).

Luke 8:51-55 (TNIV). "When he arrived at the house of Jairus, he did not let anyone go in with him except Peter, John and James, and the child's father and mother. Meanwhile, all the people were wailing and mourning for her. 'Stop wailing,' Jesus said. 'She is not <u>dead but asleep</u>.' They laughed at him, knowing that she was dead. But he took her by the hand and said, 'My child, get up!' Her <u>spirit</u> returned, and at once she stood up. Then Jesus told them to give her something to eat."

Luke 23:46 (TNIV) "Jesus called out with a loud voice, 'Father, into your hands I commit my <u>spirit</u>.' When he had said this, he breathed his last."

Hypothetical Statement 24: *Man is a body-soul-spirit being during his temporal life.*

Without his soul man does not exist, and without a body he does not exist. At the same time, without the (Field) he would not know he exists, and his body would die, because his atemporal soul (Particle), in the Sequential State, could not communicate with his body in the Temporal State.

Hypothetical Statement 25: *At the moment of the medically-defined death of an individual, which occurs when Particle/neuron two-way communications are interrupted (for whatever reason), an atemporal, Sequential-State body is generated by the Particle.*

Hypothetical Statement 26: *The Sequential-State body/soul unity is the "Spirit" of the individual.*

Hypothetical Statement 27: *The Spirit leaves the Temporal-State body entering the Out-of-Body Stage of the "dying process."*

Hypothetical Statement 28: *If the cause for Particle/neuron two-way communication interruption is reversed, the Spirit can return from the Out-of-Body Stage and the Temporal-State body is again animated by the Particle (resuscitation).*

Hypothetical Statement 29: *The next stage of the dying process involves the Spirit being transported to the Sequential State through the Field. This is the Near-Death Stage of the dying process.*

Hypothetical Statement 30: *If the cause for Particle/neuron two-way communication interruption is reversed, the Spirit can still return through the Field from the Near-Death Stage and the Temporal-State body is again animated by the Particle (resuscitation).*

Surgeon Bernie Siegel gives this account in his book, *Love, Medicine and Miracles*: "Once, as I finished a difficult emergency abdominal operation on a young, very obese man, his heart stopped just as we were about to remove him to the recovery room. He didn't respond to resuscitation. The anesthesiologist had given up and was walking out the door when I spoke out loud into the room, 'Harry, it's not your time. Come on back.' At once the cardiogram began to show electrical activity, and the man ultimately recovered fully. I can't prove it, of course, but I'm sure the verbal message made the difference. I know the experience made believers out of the other staff members who were present."

Hypothetical Statement 31: *If the cause for the Particle/neuron two-way communication interruption is not reversed and the corruption[14] of the Temporal-State body begins, the Spirit cannot return and the death process is irreversible.*

Hypothetical Statement 32: *When the dying process is complete, the individual, as a spirit, goes to God's presence – the Eternal Stage.* "Father, into thy hands I commend my Spirit." Luke 23:46

Hypothetical Statement 33: *The dying process, therefore, is a Transition from the Temporal State to the Sequential State.* Man is, thereby, "restored to life."

Hypothetical Statement 34: *With Transition, all faculties are also restored, including full access to all personal and ancestral memories.*

Man in the Sequential State will have access to all of the memories of his ancestors[15], from his parents back to Adam, the first man. He will also have access to all of the knowledge that God has chosen to share with him.

Hypothetical Statement 35: *Restored to life with an atemporal body, man will experience perfect, unaided communications between his soul and body, since both soul and body will exist in the Sequential State.*

---

[14] Decomposition of cells
[15] This refers only to events that occurred prior to the conception of the next in line.

1 Corinthians 15:42-44 (TNIV) "So will it be with the resurrection of the dead. The body is sown perishable; it is raised imperishable; it is sown in dishonor; it is raised in glory; it is sown in weakness; it is raised in power; it is sown a natural body; it is raised a spiritual body."

Conclusions

Conclusion regarding Dr. Francis Crick's "Astonishing Hypothesis"

In some parts of Dr. Crick's book, *The Astonishing Hypothesis*, he seems to express his amazement about the construction of the brain when compared to that of a digital computer. He also recognizes the relative difference in operating speed (many millions of operations per second for computers compared to "in the region of only 100 spikes per second" for a neuron) and the problem this presents to how man sees, thinks and recalls memories.

Dr. Crick says, "A brain does not look even a little bit like a general purpose computer. Different parts of the brain, even different parts of the neocortex, specialize, at least to some extent, in handling different sorts of information. Most memory appears to be stored in the very same locations that carry out current operations."

Hawkins tells his readers that the neocortex is about 2 millimeters (.079 inches) thick and, stretched flat, is roughly the size of a large dinner napkin and that no one knows precisely how many cells it contains. Some anatomists have estimated that the human neocortex contains around 30 billion neurons, but Jeff Hawkins believes "those 30 billion neurons are you." He adds, "After twenty-five years of thinking about brains, I still find this fact astounding."

Hawkins agrees again with Crick. In *On Intelligence* he says, ". . . the brain's architecture has a great deal to tell us how the brain really works and why it is fundamentally different from a computer."

The *Truly* Astonishing Hypothesis explains the apparent speed of the brain (millions of times faster than the digital computer), and suggests how the specialized parts of the brain are all interconnected (through the atemporal Particle), and also why most memory appears to be stored in the very same locations that carry out current operations (memories are efficiently stored in the Particle and are returned to the same neuron-group that transferred them).

> **"The sensual and spiritual are linked together by a mysterious bond, sensed by our emotions, though hidden from our eyes. " - Karl Wilhelm Von Humboldt (1767-1835)**

Dr. Crick's explanation for the unusual structure of the brain is as follows: "While a computer has been deliberately designed by engineers, the brain has evolved by natural selection over many, many generations of animals. This tends to produce a radically different style of design."

I am a design engineer, not a physicist and biochemist as Dr. Crick is. Here I can speak with greater authority. If the brain functions in a far, far superior manner than a computer, as we know it does, it is only logical to believe its design is superior, not just radically different. I attribute the superior design to a superior Designer. Dr.

Crick attributes the "radically different style of design" to "natural selection over many, many generations of animals." He says this but yet cannot explain the evolution of the smallest component of the brain, the neuron.

Dr. Crick calls on the scientific community to prove his hypothesis. So do I call on the scientific community to test my Hypothesis. And, borrowing his words, "If the scientific facts (gathered by the scientific community) are sufficiently striking and well established, and if they support the *Truly* Astonishing Hypothesis, then" man will have proven the existence of the human soul, a discovery that will surpass in importance the discovery of the structure of DNA. At the same time, it will reinforce "the beliefs of billions of human beings alive today."

Conclusion regarding Jeff Hawkin's *On Intelligence.*

Hawkins says, "Intelligence is measured by the capacity to remember and (compare) patterns in the world, including language, mathematics, physical properties of objects and social situations. Your brain receives patterns from the outside world, stores them as memories, and makes (comparisons) by combining what it has seen before and what is happening now." He adds in his own words, "The recalled memory is compared with the sensory input stream."

Hawkins could have written and supported The *Truly* Astonishing Hypothesis better than I. He seems to be missing only one important component - the Particle - without which his entire theory is unworkable.

Final Comment

To the scientist who believes in God, His Son, and the Holy Spirit I say the following: I have read some of your books defending your belief in a Creator. I have attended some of the conferences in which you called Darwin's theory of natural selection preposterous, when it comes to the evolution of the cell. I listened to some of your talks about the "fingerprints of God" on the creation of the human cell, but when I asked about the human soul you answered, "I don't know how the soul figures into all of this." Did God forget to tell you, or did you forget to ask? Once you know, the world needs to hear from you.

Bibliography

Barbour, Julian, *The End of Time*, The Next Revolution in
    Physics, 1999
Behe, Michael J., *Darwin's Black Box*, The Biochemical
    Challenge to Evolution, 1996
Beisswenger, Kai, *Zeitpuzzle*, a novel, 2002
Beveridge, W.I.B., *The Art of Scientific Investigation*, An
    Entirely Fresh Approach to the Intellectual
    Adventure of Scientific Research, 1950
Brown, Warren S. and Murphy, Nancy and Malony, H.
    Newton, *Whatever Happened to the Soul*, Scientific
    and Theological Portraits of Human Nature, 1998
Chopra, Deepak, M.D., *Ageless Body, Timeless Mind*, The
    Quantum Alternative to Growing Old, 1993
Crick, H.C. Francis, *The Astonishing Hypothesis*, The
    Scientific Search for the Soul, 1995
Davies, Paul, *About Time*, Einstein's Unfinished Revolution,
    1995
Davies, Paul, *The Mind of God*, The Scientific Basis for a
    Rational World, 1993
Edwards, Robert and Steptoe, Patrick, *A Matter of Life*, The
    Story of a Medical Breakthrough, 1980
Einstein, Albert, *Out of My Later Years*, 1950, 1990 Edition
Faid, Robert W., *A Scientific Approach to Christianity*, New
    Evidence Supports the Bible!, 1982
Fischbach, Gerald D., *How Neurons Communicate*,
    Scientific American, September 1992.
Hawkins, Jeff with Sandra Blakeslee, *On Intelligence*, How a
    New Understanding of the Brain Will Lead to the
    Creation of Truly Intelligent Machines, 2004
Landau, David, *Death is Not Always the Winner*, a novel,
    2001
Levine, Arnold J., *Viruses*, 1992

McInerny, Ralph M. Ph.D., *Miracles*, A Catholic View, 1986

McTaggart, Lynne, *The Field*, The Quest for the Secret Force of the Universe, 2002

Miller, Joseph D., Ph.D. *The Prospects for a Quantum Neurobiology*, 1997

Mills, Dr. Randell L., *The Grand Unified Theory of Classical Quantumm Mechanics*.

Milton, Richard, *Shattering the Myths of Darwinism*, 1997

Moody, Raymond A., M.D., *The Light Beyond*, New Explorations by the Author of *Life After Death*, 1988

Pearsall, Paul, Ph.D., *The Heart's Code*, Tapping the Wisdom and Power of Our Heart Energy, 1998

Prince, Nicholas H.E., *Are Memories Really Stored in the Brain?*, 2003

Schein, Elyse and Bernstein, Paula, *Identical Strangers*, A Memoir of Twins Separated and Reunited, 2008

Schwartz, Gary E., Ph.D., *The Afterlife Experiments*, Breakthrough Scientific Evidence of Life After Death, 2001

Ratzinger, Joseph Cardinal, Interdicasterial Commission for the *Catechism of the Catholic Church*, 1994

Rensberger, Boyce, *Life Itself*, Exploring the Realm of the Living Cell, 1995

Ross, Hugh, Ph.D., *Creation and Time*, A Biblical and Scientific Perspective on the Creation-Date Controversy, 1994

Ross, Hugh, Ph.D., *The Creator and the Cosmos*, How the Greatest Scientific Discoveries of the Century Reveal God, 1993

Sabom, Michael, M.D., *Light & Death*, One Doctor's Fascinating Account of Near Death Experiences, 1998

Schroeder, Gerald L., *Genesis and the Big Bang*, The Discovery of Harmony Between Modern Science and the Bible, 1992

Schwartz, Gary E., Ph.D. and Chopra, Deepak, M.D., *Science and Soul*, The Survival of Consciousness After Death, a tape dialog, 2001

Sheldrake, Rupert, Ph.D., *A New Science of Life,* The Hypothesis of Morphic Resonance, 1995

Siegel, Bernie, M.D., *Love, Medicine and Miracles,* Lessons Learned About Self-Healing from a Surgeon's Experience with Exceptional Patients, 1986

Thorne, Kip S., *Black Holes & Time Warps*, Einstein's Outrageous Legacy, 1994

Ward, Keith, *In Defence of the Soul*, 1998

Wigglesworth, Smith, *Healing*, Experience God's Miracles, 1982

Other references can be found in the body of the text and the following Index.

Index

A New Science of Life, 59
action potentials, 22, 23, 43
adequate sleep, 45
aging, 44, 45
ancestral memories, 11, 52
Are Memories Really Stored in the Brain, 28, 58
association areas, 25
Astonishing Hypothesis, 7, 8, 53, 57
atemporal, 11, 12, 20, 43, 51, 52, 54, 66, 74
atemporally, 11, 29
atheist, 11, 13
atheistic, 12
Atwater, P.H.M, 48
axon, 22, 23
Ball, John, 34
benign childhood epilepsy, 25
Bible, 19, 47, 49, 57, 58, 69, 70, 71, 72
binding problem, 20
biological, 9, 13, 20, 31, 44
brain, 8, 12, 19, 20, 23, 24, 25, 26, 27, 28, 29, 30, 31, 32, 33, 34, 46, 48, 53, 54, 55, 57, 58, 75
Breath of God, 21
Buehler, Guenter Albrecht, 35, 67
cancer, 38, 39, 41, 42, 46
cancer cell, 39, 41, 42
cell division, 36, 37, 38, 39, 40, 42
cell-specific instructions, 39, 41, 42, 43
cellular phone, 46
centrosome, 40
chromosome(s), 24, 36, 39, 40, 41, 42
coded genetic traits, 23
comparisons, 30, 55

comprehensible, 26, 33
computer, 31, 32, 53, 54, 79
conception, 14, 36, 52, 75
Concurrent, 16, 17, 36
Concurrent State, 17, 18
consciousness, 9, 28, 30, 43, 58, 67, 78
control and recall process, 32
Corresponding, 39, 42
corruption, 52
cortex, 11, 21, 25, 26, 30, 34
cosmos, 9, 13, 58
creation, 11, 15, 56, 57
Creation, 13, 58
Creator, 14, 15, 55, 58
Crick, Francis, 4, 7, 8, 9, 12, 13, 14, 19, 20, 21, 26, 35, 47,
    53, 54, 55
Darwin, 55, 57
Darwinian, 13
death, 7, 15, 43, 47, 48, 49, 50, 51, 52, 57, 58
dendrites, 22, 25
determinants of longevity, 45
digital computer, 31, 53, 54
disturbance, 24, 25, 43, 46
division limits, 39
DNA, 7, 13, 14, 20, 23, 24, 35, 36, 39, 40, 41, 42, 45, 46, 55,
    67, 76
DNA polymerase molecule, 41, 42, 45
dying process, 51, 52
electromagnetic fields, 46, 75
electromagnetic radiation, 42
environmental factors, 36
epithelial cells, 43
Erlykin and Wolfendale, 45
Eternal Stage, 52
evolved, 13, 54, 67
existence, 8, 13, 16, 46, 47, 55, 72
Exodus, 17, 67, 69, 72

exonuclease, 41, 42
extemporaneous discussions, 34
extemporaneous speech, 31
Field, 21, 24, 25, 26, 27, 30, 36, 39, 41, 42, 43, 46, 51, 58
fingerprints of God, 56
Fischbach, 22, 24, 57
genome, 35
Gilman, Robert, 28
Glover, David, 40
God, 11, 12, 13, 15, 17, 19, 21, 36, 49, 52, 55, 56, 58, 59, 67, 68, 69, 70, 71, 72, 73
Grolier Multimedia Encyclopedia, 45
growth, 36, 39, 41, 76
Hardin, Garrett, 35, 67
Hawking, Stephen, 9
Hawkins, Jeff, 4, 8, 9, 10, 11, 12, 19, 23, 25, 26, 30, 31, 32, 33, 34, 47, 53, 54, 55, 57
Hayflick, Leonard, 44, 45
healing, 15, 36, 39, 43, 59, 75, 76
hereditary traits, 20, 36
homunculus, 35, 75
human, 5, 7, 8, 13, 14, 20, 31, 35, 36, 37, 38, 44, 45, 54, 55, 56, 57, 67, 74, 75
Humboldt, Karl, 54
hypothetical statements, 16, 20, 24, 26, 29, 30, 49, 73, 75, 76
incomprehensible, 25, 33
intelligence, 8, 9, 10, 12, 54, 55, 57
intuition, 11, 75
Jacob's Ladder, 19
Lazarus, 50
life, 15, 20, 35, 21, 36, 45, 49, 51, 52, 57, 58, 59, 73, 74
Life Force, 8
life-limiting radiation, 45
Likeness, 35, 36, 39, 46, 73
Love, Medicine and Miracles, 50, 59
low-level electromagnetic radiation, 46
memories, 7, 10, 11, 28, 29, 30, 31, 33, 52, 53, 54, 55, 58, 75

memory, 11, 14, 28, 29, 32, 33, 34, 46, 53, 54, 55, 75
microtubules, 40
Mills, Dr. Randell L., 58, 66
mind, 6, 8, 10, 11, 13, 14, 15, 21, 27, 43, 49, 57
mitosis, 40, 41, 42
natural selection, 13, 44, 54, 55
Near-Death Stage, 51
neocortex, 11, 21, 29, 32, 33, 53, 54, 74
nerve cells, 7
nervous system, 21, 43, 66
neuorscience, 9
neuro-network, 30
neuron(s), 5, 7, 11, 19, 20, 21, 22, 23, 24, 25, 26, 27, 29, 30,
    31, 32, 33, 43, 46, 51, 52, 53, 54, 55, 57
neurotransmitter, 23, 66
New Scientist, 29
nucleosome assembly, 24
nucleosomes, 23, 24
nucleotides, 41
nucleus, 24, 36, 42
On Intelligence, 10, 12, 54, 55, 57
one hundred-step rule, 31
operating speed, 53
Out-of-Body experience, 48
Out-of-Body Stage, 51
Parallata, 25, 26, 29, 30, 32
Parallatum, 25
parallel computer, 31, 32
Particle, 11, 20, 21, 24, 25, 26, 27, 29, 30, 32, 33, 34, 35, 36,
    39, 40, 41, 42, 43, 45, 46, 51, 52, 54, 55, 74, 77
Pearsall, Paul, 28, 58
percepts, 33
pericentriolar material, 40
physical regeneration, 39
Plato, 10, 11, 12
polymerase, 41, 42, 45
postsynaptic membrane, 23

postsynaptic neuron, 23
power lines, 46
predictions, 30
prefrontal cortex, 21
primary auditory area - A1, 25
primary sensory areas, 25
primary somatosensory region - S1, 25
primary visual area - V1, 25
Prince, Nicholas, 28, 58
processes of aging, 45
procreation, 36
pyramidal cell, 32
pyramidal neuron(s), 21
quantum physics, 12, 66
radiation, 42, 45, 46
Reacholi, Michael, 46
reality, 10, 11, 12, 67
religion, 7, 11
Religious leaders, 19
REM sleep, 43, 45, 77
repository, 33
reproduction, 15, 45, 46
resurrection, 49, 52
resuscitation, 50, 51
retention, 28
returned from death's door, 47
Reynolds, Pam, 48, 77
RNA, 45
Ross, Hugh, 45, 58
Sabom, Michael, 47, 48, 58
Scientific American, 22, 40, 57
scientist(s), 8, 9, 12, 13, 14, 29, 34, 47, 55, 75
sequence of patterns, 30
Sequential, 16, 17, 18, 20, 21, 25, 26, 27, 30, 51, 52
Sequential State, 16, 17, 18, 20, 21, 51, 52
sequentially, 17, 29, 30, 33
Sequential-State body/soul unity, 51

Sheldrake, Rupert, 59, 74, 75
Siegel, Bernie, 50, 59
singularity, 20
sleep, 43, 50, 45
Somehow, 42, 75
soul, 7, 8, 9, 10, 11, 19, 20, 26, 27, 30, 49, 51, 52, 55, 56, 57, 58, 59, 73, 74, 76
spinal reflexes, 32
spirit, 7, 49, 50, 51
Spirit, 51, 52, 55
Standstill, 48
supernovae, 45
Swezey, Laurie, RN, 77
synapses, 22, 23, 25, 31, 32
temporal, 16, 20, 26, 27, 29, 32, 36, 43, 45, 51, 52
Temporal State, 16, 17, 18, 21, 24, 25, 26, 51, 52
The Heart's Code, 28, 58
Theistic Paradigm for Biology, 19
Thinking Solutions, 34
Tillis, Ray, 29
transition, 47, 52
transplant recipients, 28
Truly Astonishing Hypothesis, 6, 7, 9, 13, 14, 23, 54, 55
unity, 20, 27, 30, 51
universe, 12, 58
Vela supernova, 45
vision, 27
WHO's Electromagnetic Fields Project, 46
Wilcox, David, 19, 20
zygote, 13, 14, 35, 36, 67

End Notes

---

[i]   "Experiments by the early part of the 20th century had revealed that both light and electrons behave as waves in certain instances and as particles in others. This was unanticipated from preconceptions about the nature of light and the electron. Early 20th century theoreticians proclaimed that light and atomic particles have a 'wave-particle duality' that was unlike anything in our common-day experience. The wave-particle duality is the central mystery of the presently accepted atomic model, quantum mechanics, the one to which all other mysteries could ultimately be reduced." Dr. Randell L. Mills, in his book, *The Grand Unified Theory of Classical Quantum Mechanics.*

Is Dr. Mills suggesting that all matter has a dual nature, one physical and one atemporal and that understanding the dual nature of all matter will lead to . . . ?

"There must be a point in this reductionist program where molecular biology enters the domain of quantum physics, a point at which classical, Newtonian, deterministic theory (the usually unacknowledged underpinning of modern biology) must give way to quantum mechanical interpretation. Nowhere will this be seen more clearly than in attempts to understand the mechanism and function of the central nervous system and the diseases to which it is prone. Already, quantum mechanical considerations are necessary in the modeling of neurotransmitter receptor structure (Pardo et al., 1996). It is possible that some of the "hard" problems in neurobiology, such as the nature, origin, and development of conscious experience will require a quantum perspective. The inevitable intersection of neurobiology with quantum mechanics may lead to a twin understanding of cognition and the role of the observer in quantum mechanics. That knowledge in turn will have profound implications for psychiatric medicine, the definition of 'human', and perhaps the interpretation of physical reality itself. But to evaluate such possibilities it is necessary to briefly consider both philosophical approaches to the nature of consciousness and the essentials of quantum mechanics. The difficulties in the interpretation of quantum mechanics (as opposed to its pragmatic application) may provide unique insight into some of the most difficult problems of neurobiology."

The Prospects for a Quantum Neurobiology by Joseph D. Miller Ph.D, Department of Pharmacology, Texas Tech University Health Sciences Center, 1997.

ii "I am unable to believe that any machine (referencing the cell) can be designed that contains an instruction library (DNA) which anticipates all the mishaps and glitches of a billion years of evolution without crashing over and over again." Guenter Albrecht-Buehler, Professor of Cell and Molecular Biology at Northwestern University.

In other words, he does not believe that the human cell, containing DNA, could have evolved.

"A set of blueprints is not a house; the DNA of a zygote is not a human being." Garrett Hardin, professor of biology at the University of California at Santa Barbara
iii Let's see how God defined Himself in Exodus 3:14 (TNIV). Because there are many interpretations of Exodus 3:14 in which God tells us who He is, what He is and where He is, interpretations which in my opinion do not all represent what God actually said to Moses, I have incorporated in this end note my research regarding this all important verse. The following is the entire set of verses from Exodus 3:1 through 3:14. I have included this entire section to keep the story in proper context. The NIV translation was used.

69

¹ Now Moses was tending the flock of Jethro his father-in-law, the priest of Midian, and he led the flock to the far side of the desert and came to Horeb, the mountain of God. ² There the angel of the LORD appeared to him in flames of fire from within a bush. Moses saw that though the bush was on fire it did not burn up. ³ So Moses thought, "I will go over and see this strange sight—why the bush does not burn up."

⁴ When the LORD saw that he had gone over to look, God called to him from within the bush, "Moses! Moses!" And Moses said, "Here I am."

⁵ "Do not come any closer," God said. "Take off your sandals, for the place where you are standing is holy ground." ⁶ Then he said, "I am the God of your father, the God of Abraham, the God of Isaac and the God of Jacob." At this, Moses hid his face, because he was afraid to look at God.

⁷ The LORD said, "I have indeed seen the misery of my people in Egypt. I have heard them crying out because of their slave drivers, and I am concerned about their suffering. ⁸ So I have come down to rescue them from the hand of the Egyptians and to bring them up out of that land into a good and spacious land, a land flowing with milk and honey. ⁹ And now the cry of the Israelites has reached me, and I have seen the way the Egyptians are oppressing them. ⁷ The LORD said, "I have indeed seen the misery of my people in Egypt. I have heard them crying out because of their slave drivers, and I am concerned about their suffering. ⁸ So I have come down to rescue them from the hand of the Egyptians and to bring them up out of that land into a good and spacious land, a land flowing with milk and honey. ⁹ And now the cry of the Israelites has reached me, and I have seen the way the Egyptians are oppressing them. ¹⁰ So now, go. I am sending you to Pharaoh to bring my people the Israelites out of Egypt."

¹¹ But Moses said to God, "Who am I, that I should go to Pharaoh and bring the Israelites out of Egypt?"

¹² And God said, "I will be with you. And this will be the sign to you that it is I who have sent you: When you have brought the people out of Egypt, you will worship God on this mountain."

¹³ Moses said to God, "Suppose I go to the Israelites and say to them, 'The God of your fathers has sent me to you,' and they ask me, 'What is his name?' Then what shall I tell them?"

¹⁴ God said to Moses, "I am who I am. This is what you are to say to the Israelites: 'I AM has sent me to you.'" "I am who I am."

Is that really what God said, I wondered? Is that really how God answered Moses? Is that a proper answer to the question, "Suppose I go to the Israelites and say to them, 'The God of your fathers has sent me to you,' and they ask me, 'What is his name?' Then what shall I tell them?" and God answers, "I am who I am?" Is that what God actually said? The Bible is the word of God, but did the translators get it right from the original Hebrew? Others aren't so sure either. Here's a list of the various translations of God's answer in Exodus 3:14.

"I will exist which I will exist."	Modern Translation of Hebrew Bible
"I am who I am."	New International Version
"I will be what I will be."	New International Version (alternate)
"I am who I am."	New American Standard Bible
"I am who I am."	The Message
"I am who I am and what I am and I will be what I will be."	Amplified Bible
"I am who I am."	New Living Translation
"I will be what I will be." (alternate)	New Living Translation

"I am that I am."	King James Version
"I am who I am."	English Standard Version
"I am the eternal God."	Contemporary English Version
"I am who I am."	New King James Version
"I am who I am."	New Century Version
"I am that I am."	21st Century King James Version
"I am that I am."	American Standard Version
"I am that which I am."	Young's Literal Translation
"I am that I am."	Darby Translation
"I am who I am."	Holman Christian Standard Bible
"I am who I am."	New International Reader's Version
"I am who I am."	New International Version – UK
"I am who I am."	Today's New International Version
"I am who am."	New American Bible
"I am who am."	Douay-Rheims Version
"I am who I am."	Revised Standard Version
"I am he who is."	New Jerusalem Bible

So, of these translations, with all due respect, I ask, which would serve as a proper answer to Moses' question? Moses needed to know. After all, God had just said, "I am sending you to Pharaoh to bring my people the Israelites out of Egypt." Moses was a shepherd. He was overwhelmed by the assignment. He needed a straight answer. If you were Moses, would "I am who I am?" be a good, reasonable, loving answer? Not for me. If I was asked by the publisher of this book, "Who should I tell the readers you are?" and I answered, "I am who I am." Would that have been a proper response? Let's try eliminating that interpretation from the list and see what choices remain.

"I will exist which I will exist."	Modern Translation of Hebrew Bible
"I will be what I will be."	New International Version

(alternate)

"I am who I am and what I am and I will be what I will be."
Amplified Bible

"I will be what I will be."   New   Living   Translation
(alternate)

"I am that I am."   King James Version

"I am the eternal God."   Contemporary   English
Version

"I am that I am."   21st Century King James
Version

"I am that I am."   American Standard Version

"I am that which I am."   Young's Literal Translation

"I am who am."   New American Bible

"I am that I am."   Darby Translation

"I am who am."   Douay-Rheims Version

"I am he who is."   New Jerusalem Bible

Now we have a shorter list of interpretations of what
God said to Moses. How about, "I am that I am?" Still
doesn't answer the question as Moses needed it answered.

"I will exist which I will exist."   Modern Translation of
Hebrew Bible

"I will be what I will be."   New International Version

"I am who I am and what I am and I will be what I will be."
Amplified Bible

"I will be what I will be."   New Living Translation

"I am the eternal God."   Contemporary   English
Version

"I am that which I am."   Young's Literal Translation

"I am who am."   New American Bible

"I am who am."   Douay-Rheims Version

"I am he who is."   New Jerusalem Bible

We're down to 9 interpretations. How about, "I will
exist which I will exist?" Nope. Or, "I will be what I will
be?" Not that one either.

Then there's, "I am who I am and what I am and I
will be what I will be?" Moses might have asked, "Would
you say that again?" And also, "I am that which I am." Not

any better.

"I am the eternal God."	Contemporary English Version
"I am who am."	New American Bible
"I am who am."	Douay-Rheims Version
"I am he who is."	New Jerusalem Bible

That leaves us with three good answers from four interpretations, "I am the eternal God," "I am who am" and "I am he who is." Well, "I am the eternal God" is certainly a great answer, but it is not a translation; it is a dynamic equivalent of what was said by God.

That leaves us with two interpretations which have the same meaning. God is the only being in existence who simply is. For him, everything is in the present.

Remember, now, God finished his answer with, "This is what you are to say to the Israelites: 'I AM has sent me to you.' " "I AM," therefore provided Moses with a clear statement of who was sending him.

Here is the Hebrew text of Exodus 3:14

לארשי ינבל רמאת הכ ,רמאיו ;היהא רשא היהא ,השמ-לא םיהלא רמאיו **די**,
.היהא ,מכילא ינחלש

It reads from right to left. This is the subject phrase:

היהא רשא היהא

It translates literally, "Am who Am." I conclude, "I am who am." is the correct translation. It is a wonderful and complete, no, a <u>perfect</u> statement of who our God is.
[iv] C. S. Lewis once wrote something like this, "We are so little reconciled to time that we are astonished by the passage of it. We are like a fish which is constantly surprised by the wetness of water, as if it was destined to become a land animal." What he meant, of course, was that we are surrounded by time like the fish is with water and preoccupied with it because we are, in fact, destined to

become timeless beings.

v   God is not limited by time or sequence (Hypothetical Statement #1). Jesus said himself, "Before Abraham was born, I am." John 8:58 (TNIV) In Apostle John's gospel, John 1:1-4 (TNIV), I recently read again, "In the beginning was the Word, and the Word was with God, and the Word was God. He was in the beginning with God. All things came to be through him, and without him nothing came to be. What came to be through him was life . . ." I stopped there. My attention was called to Genesis 1:26. God said, "Let us make man in our image, in our likeness . . ." I looked up the Hebrew for the word "image" and found the synonym "form." I looked up the Hebrew for "likeness" and found "resemblance." Of what person in the Trinity do we have the same likeness, resemblance, image and form? I've concluded, the Lord Jesus was there when we were created. Genesis 2:7 says, "The Lord God formed man out of the clay of the ground and blew into his nostrils the breath of life, and so man became a living being." It was Jesus that formed us and it was Jesus who gave us life. Yet it was Jesus who was born of the virgin Mary, having come as the Savior of the world. God is not limited by time or sequence.

vi   "But what *is* the Soul? Men have struggled over this question for centuries because they have tried to prove, using the Scriptures, an unscriptural idea. As a result their definitions of what the 'soul' is have been undefined, vague, and elusive. It has been taught that the soul is 'something' in us, but no one seems able to explain either where or what it is. Theologians like to claim that this vague unknown entity is the real intelligent being and that the body which 'houses' it is just some sort of metaphysical tool. As science probes into the processes by which our bodies operate, man is finding that we are little more than a rather extensive chemical and electrical factory and that the workings of the physical affect the function of the mental and emotional and vice versa. This seems directly at odds with the way a Methodist Bishop of some years ago described the soul: 'It is without interior or exterior without body, shape or parts, and you could put a million of them into a nutshell.' This well meant attempt to describe the soul seems to us rather a good definition of nothing! But our question remains unanswered. Merely scoffing at false answers is no help because there remain aspects of humanity which do defy description."

The Herald of Christ's Kingdom is the official publication of the Pastoral Bible Institute. Published since 1918.

vii My search for a definition of the "fabric of the soul" may not have been an exhaustive one, but I found only a few poets who have used it to express the "heart" of the soul, which, of course, they did not define. Let the "fabric of the soul" be defined here, therefore, as a descriptor for the fact that the Particle exists at the functional center of each of the billions of cells of our bodies and yet it is, at once, a single atemporal Particle, unique to each of us – resulting in a spiritual fabric with a function not unlike but greater than the neocortex.

viii "Morphogenetic Fields carry information only (no energy) and are available throughout time and space." Dr. Rupert Sheldrake, Author, *A New Science of Life*

> Morphogenetic: relating to or concerned with the development of normal organic form.
>
> Merriam-Webster

Some scientists who were looking for an explanation of everything, and especially how so much information could exist in a single cell at conception, have considered that the information necessary to develop the (body) of a new individual must come from a field that is surrounding the primary (stem) cell - a field that somehow has all of the information necessary and specific to the individual. They call it the morphogenetic field. They also believe it pervades all space, interacts with all matter and energy, and is even the basis of the Unifying Field Einstein was searching for. Some have suggested that the morphogenetic field interfaces with the electromagnetic field of the brain (not far off) and is involved in how we can recall memories (right on). Dr. Sheldrake goes on to say, "Morphogenetic fields are basically non-physical (spiritual?) blueprints that give birth to forms."

[ix] Do you remember playing on a freshly cut lawn? Did you ever run your hand over the top of the grass? Remember the feeling you got on the palm of your hand? Extend out your arm right now with the palm of your hand facing down.
Move it back and forth and think about how it felt when you touched the grass. Can you feel the blades of grass touching your palm? How does that happen?

[x] "I myself find it difficult at times to avoid the idea of a homunculus. One slips into it so easily." Dr. Francis Crick's intuition may have been suggesting that there is a spiritual aspect to the human.

[xi] I read in an article on the process of healing that fibroblasts are cells key to the process. Quoted from the accompanying article: "The principle function of fibroblasts is to maintain the structural integrity of connective tissues. Fibroblasts have diverse appearances depending on their location and activity." Interesting. The article continues, "Fibroblasts can often (somehow) retain positional memory of the location and tissue where they had previously resided. It is fibroblasts and related connective tissues which sculpt the 'bulk' of an (injured) organism." Look at Hypothetical Statement 5! Fibrocytes are newly identified cells that can rapidly and specifically migrate into the site of tissue injury at the time of injury. They are derived from bone marrow.
Just what directs and controls the actions of fibroblasts and fibrocytes? DNA is not even mentioned in this article.

Could it be the soul as suggested by Hypothetical Statement 5? I think so.

Quotations above from C. Michael Gibson M.S., M.D., Associate Professor of Medicine, Harvard Medical School.
[xii]   "Moist wound healing is the name given to the observation that a wound that is kept optimally moist will have better outcomes than one that is allowed to dry out. Studies have shown that a moist wound heals between three and five times faster than a dry wound.

The idea of moist wound healing was first defined during the 1960s. During this time, early pre-clinical and clinical research conducted by the British pioneer, George D Winter, first demonstrated the benefit of a moist environment in optimizing wound healing.

Our understanding of moist wound healing has come a long way since the days of the first research of Dr Winter. We now understand why wound healing is promoted by a moist environment. This is due to several, parallel processes. Firstly, by preventing scab or crust formation over the wound bed, a moist wound environment eliminates the energy and time that would have been required for the body to breakdown these materials. Keratinocyte-travel time and distance across the wound surface are also greatly reduced, as the cells are able to easily migrate across the moist wound bed rather than burrow underneath the wound bed to find a moist area upon which to move forward. A moist environment also traps enzymes within the wound bed, facilitating autolytic debridement. And finally, a moist wound environment preserves growth factors within the wound fluid, and increases fibroblast proliferation and collagen synthesis." (see also End Note [x])

Posted on Wound Education Social Network by Laurie Swezey, RN, BSN, CWS,CWOCN's.
[xiii]   Someone close to me called the other day to tell me he

was diagnosed with myelofibrosis. Some, not all, of the stem cells of his bone marrow have mutated to divide into cells that produce excessive proteins consisting of tiny fibers. That's the "fibrosis" part. Lower red blood cell production in his bone marrow resulted. Some of the blood cell production was picked up by his spleen, which caused it to swell. The problem, in my non-medical opinion, is that errant stem cells in his bone marrow are having difficulty making division decisions regarding the type of cells to become – poor communications with the Particle. I told him to ask his doctor if he can make the mutant cells healthy. The doctor gave him hydroxyurea instead, a drug which kills fast dividing cells (and some normal cells).

[xiv] "Do the eyes scan dream images during rapid eye movement sleep? Evidence from the rapid eye movement sleep behaviour disorder model." A study by Laurène Leclair-Visonneau, Delphine Oudiette, Bertrand Gaymard, Smaranda Leu-Semenescu and Isabelle Arnulf, published in Brain, a Journal of Neurology – May 16, 2010.

[xv] **Pam Reynolds**

"I remember seeing several things in the operating room when I was looking down (from above the operating table). The saw (the surgeon was using) looked like an electric toothbrush. And the saw had interchangeable blades too, but these blades were in what looked like a socket wrench case.
Someone said something about my veins and arteries being very small. I believe it was a female voice (speaking) and that it was Dr. Murray, the cardiologist. I remember thinking that I should have told here about that. There was a sensation like being pulled, but not against (my) will. I was going on my own accord because I wanted to go. It was like being taken up in a tornado vortex only you're not spinning around. At the end there was this little tiny pinpoint of light that kept getting bigger and bigger. The light was incredibly bright. I noticed that as I began to discern different figures

in the light, they began to form shapes I could recognize and understand. I could see that one of them was my grandmother. Everyone I saw, looking back on it, fit perfectly into my understanding of what that person looked like at their best during their lives. I recognized a lot of people. My uncle Gene was there; so was my great-great aunt Maggie. On (my father's) side of the family; my grandfather was there. They were specifically taking care of me, looking after me. They would not permit me to go further. It was communicated to me that if I went all the way into the light, something would happen to me physically. They would be unable to put this _me_ back into the body _me_. I wanted to go into the light, but I also wanted to come back. I had children to be reared. My uncle . . . took me back through the end of the tunnel. Everything was fine. I did want to go. But then I got to the end of (the tunnel) and saw the thing, my body. I didn't want to get into it. It looked like what it was, dead. It scared me and I didn't want to look at it. I felt a definite repelling and at the same time pulling from the body. The body was pulling and the tunnel was pushing. It was like diving into a pool of ice water . . . It hurt. When I came back (into my body), they were playing 'Hotel California' (in the operating room) and the line was, 'You can check out anytime, but you can never leave.' When I regained consciousness, I mentioned to Dr. Brown that (playing that song) was incredibly insensitive.

Acknowledgements

My primary acknowledgement goes to my beautiful wife, Kim, who not only encouraged me but asked me to write this book. Kim has frequently found for me important articles relating to the Hypothesis.

My second acknowledgement must go to my oldest son, David, who is a certified computer network administrator. He has always sent me articles relating to the Hypothesis and sent me Jeff Hawkins' book, *On Intelligence*.

My third acknowledgement goes to my pastor, Dale Shaw, senior pastor of Pequea Church. I asked him to review the Hypothesis to be sure the related theology is in line with his Christian values and beliefs.

www.ingramcontent.com/pod-product-compliance
Lightning Source LLC
Chambersburg PA
CBHW071611170526
45166CB00003B/1052